Stemmie, the Little Stem Cell™

When Rick Was Sick

By Auntielynny™

Written for Children of All Ages

The information in this book is for educational and entertainment purposes only. It is not intended to be a substitute for professional medical care. The author and publisher are not responsible for any consequences resulting from any information, procedures or errors in this book.

An Auntielynny Publication
P.O. Box 74672
San Clemente, CA 92673

Website: http://www.Auntielynny.com

Creative Director: Nancy Ratkiewich
Illustration: Christina Rosenthal
Design & Production: njr productions

ISBN_978-0-692-22416-8

To order additional copies, please contact

CreateSpace™
www.createspace.com
1-866-308-6235

This book is dedicated
to everyone wanting to live a healthier
and more compassionate life!

*Compassion helps us serve
and Wisdom helps us serve well.*

The Dalai Lama

A special thanks to
Nancy Ratkiewich and Christina Rosenthal
for all their joy, humor and wisdom
while working on this project.

Once upon a time
there was a boy named Rick.
He liked to run and play –
– but he was often sick.

He was taken to some doctors
to see what might be wrong,
but nothing seemed to help him
get well for very long.

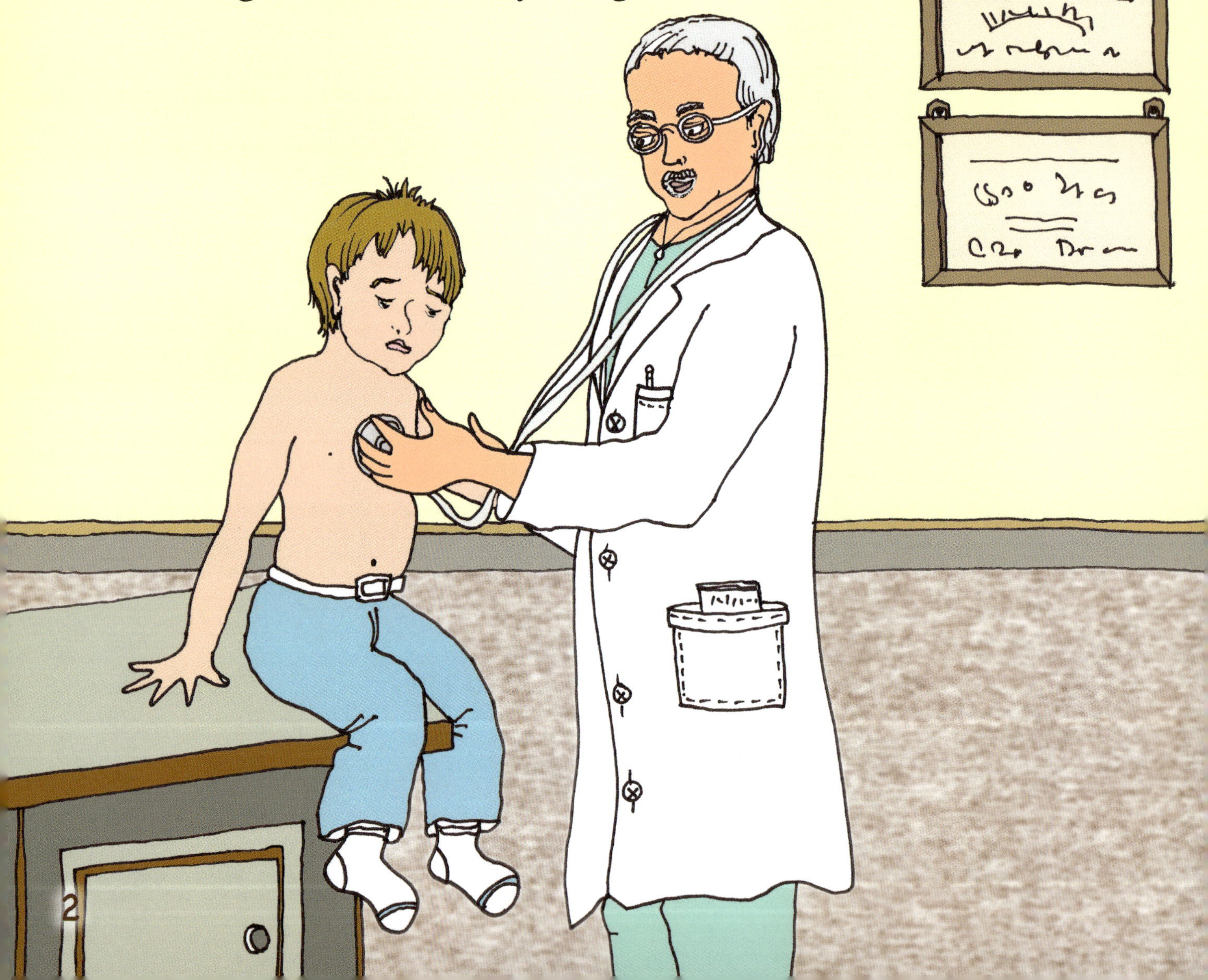

Ricky's mother was so worried.
What were they going to do?
He could seldom go to school
because he always had the flu.

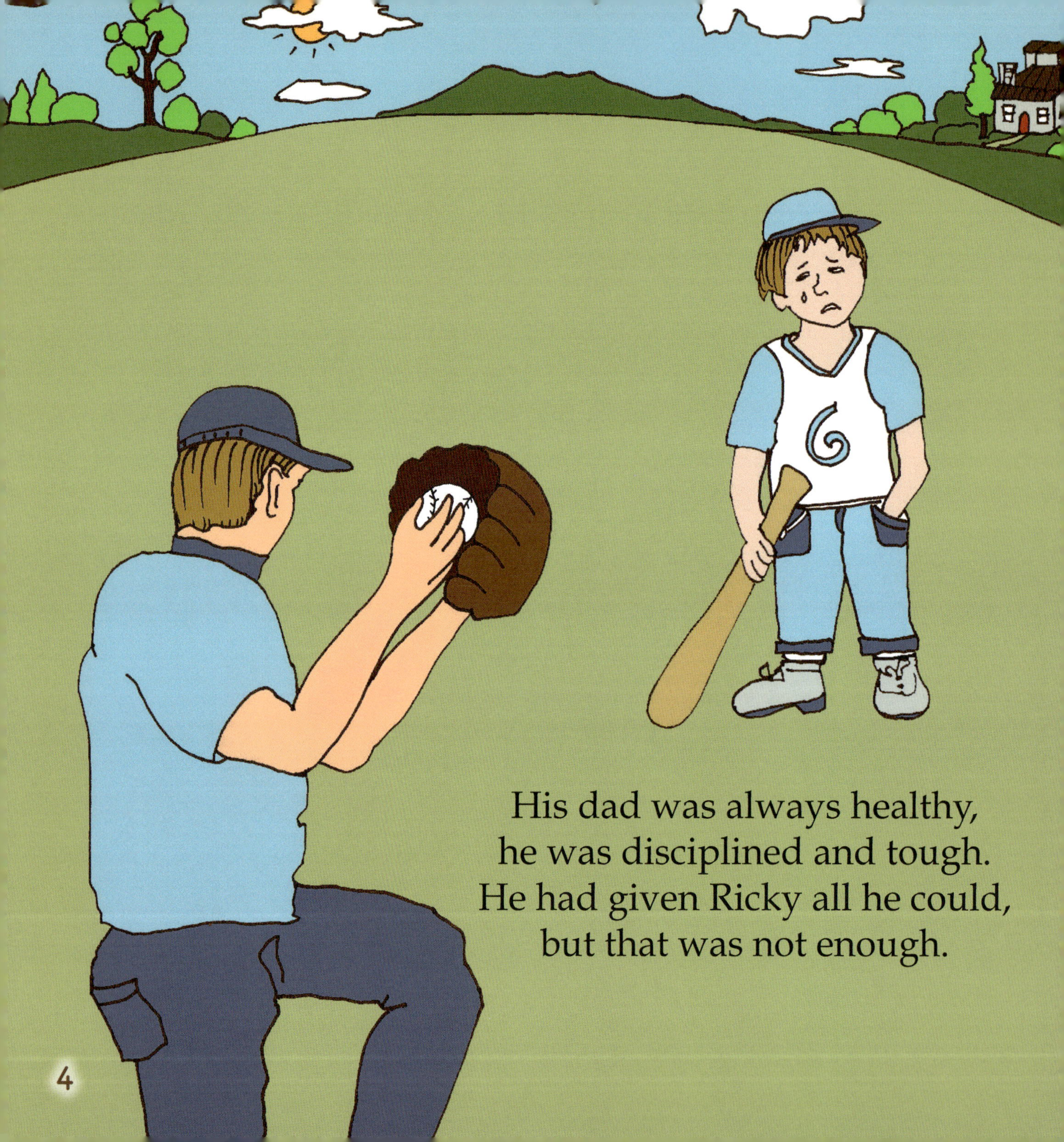

His dad was always healthy,
he was disciplined and tough.
He had given Ricky all he could,
but that was not enough.

His dad then made a solemn wish
with all his heart and mind,
that Rick would know his inner strength
and leave the past behind.

That night as Ricky tossed and turned
he had the strangest dream
about a loving stem cell
and his little wellness team.

"Some of us are memory cells,"
Stemmie seemed to say.
"We were there with you before your birth,
and help you grow each day.

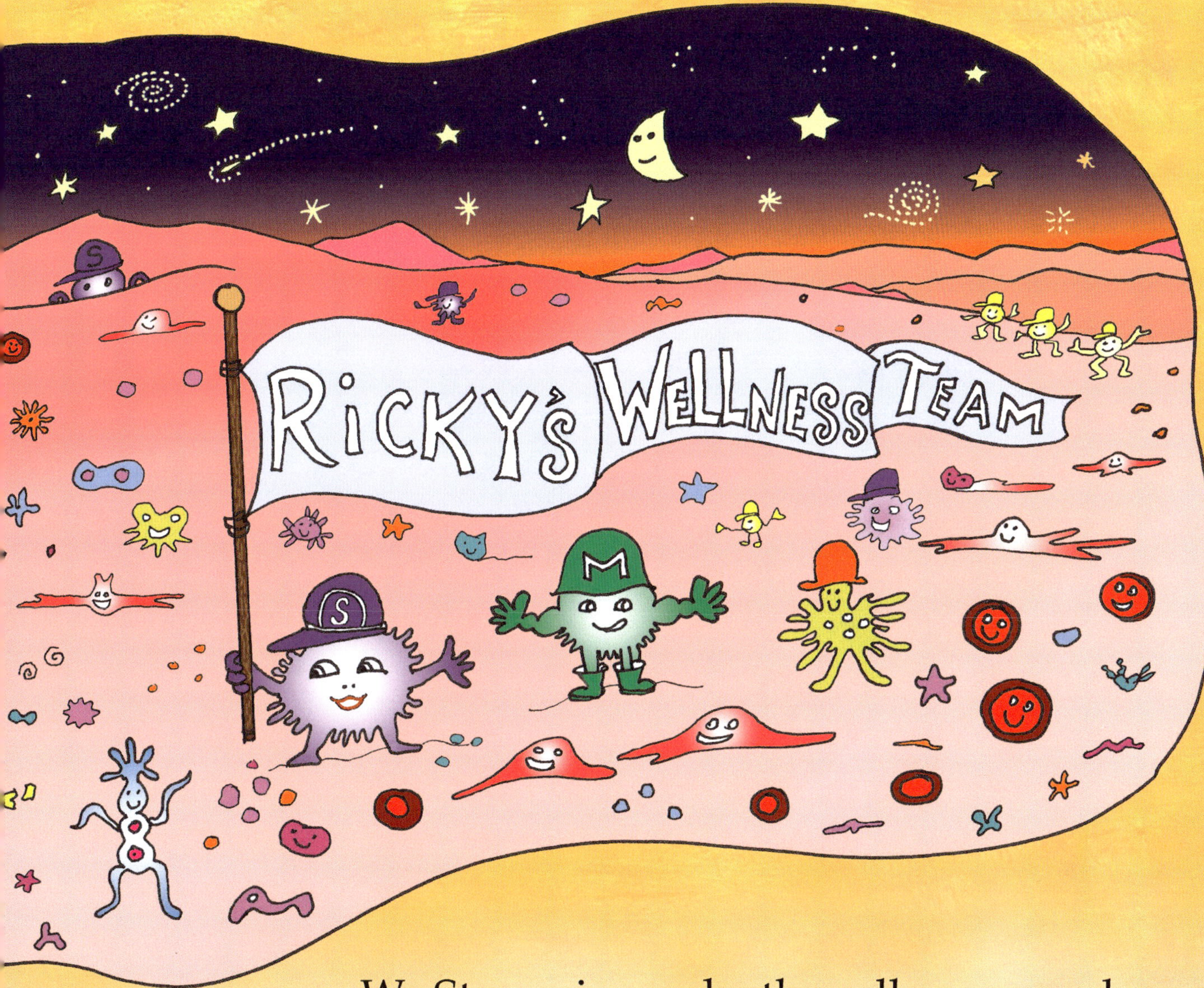

We Stemmies make the cells you need
to walk and talk and see,
and all our friends come help us,
if you fall and hurt your knee."

"I don't know if you can help me.
I'm so big and you're so small.
There would need to be a lot of you
to help me if I fall."

"We can help you if you let us,
if you show us that you care.
With tender love and exercise,
we'll be an awesome pair…

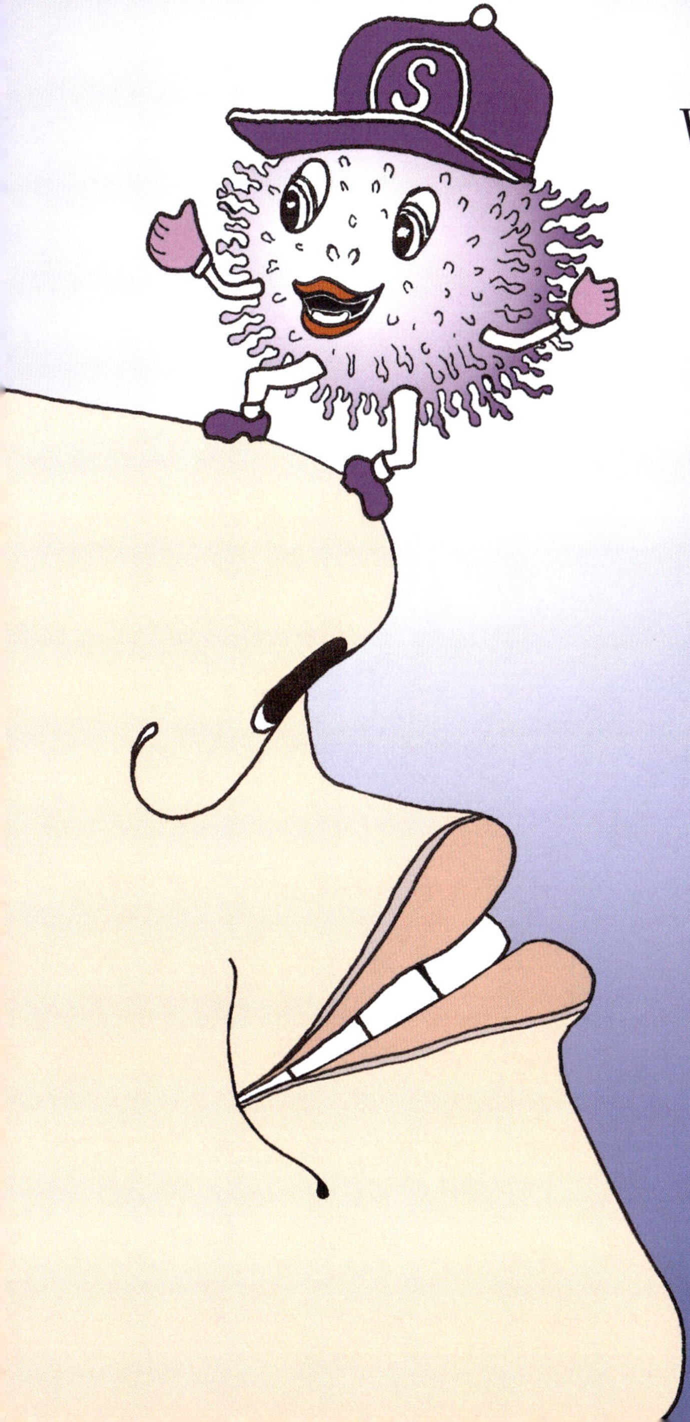

Of course…
we'd also like a good night's sleep,
inspiring thoughts, fresh foods to eat,
we'd like to have clean water and air,
plus lots of learning –
if you really care."

“Well…tell me more about you
and how you fight disease,
if you’re sure that you can help me,
it would put my mind at ease.”

"When your other cells are injured,
they're not happy and they scream,
then cells like Mac come in as
first responders on our team...

Mac sincerely tries to problem-solve,
to find the middle ground with
the bad guys hurting others
then he starts the second round.

He gobbles up the germs
who insist on causing trouble,
they drop into his stomach
then become a pile of rubble.

Then Mac sends out the Cytos
who show us where you hurt–
we come close to where the damage is
then start to do our work.

We double all our numbers
1 to 2, then 2 to 4,
4 to 8 and 8 to 16,
then we double out some more.

We divide into a million cells
then march into the zone.
We replace our brothers who have died
and make their work our own.

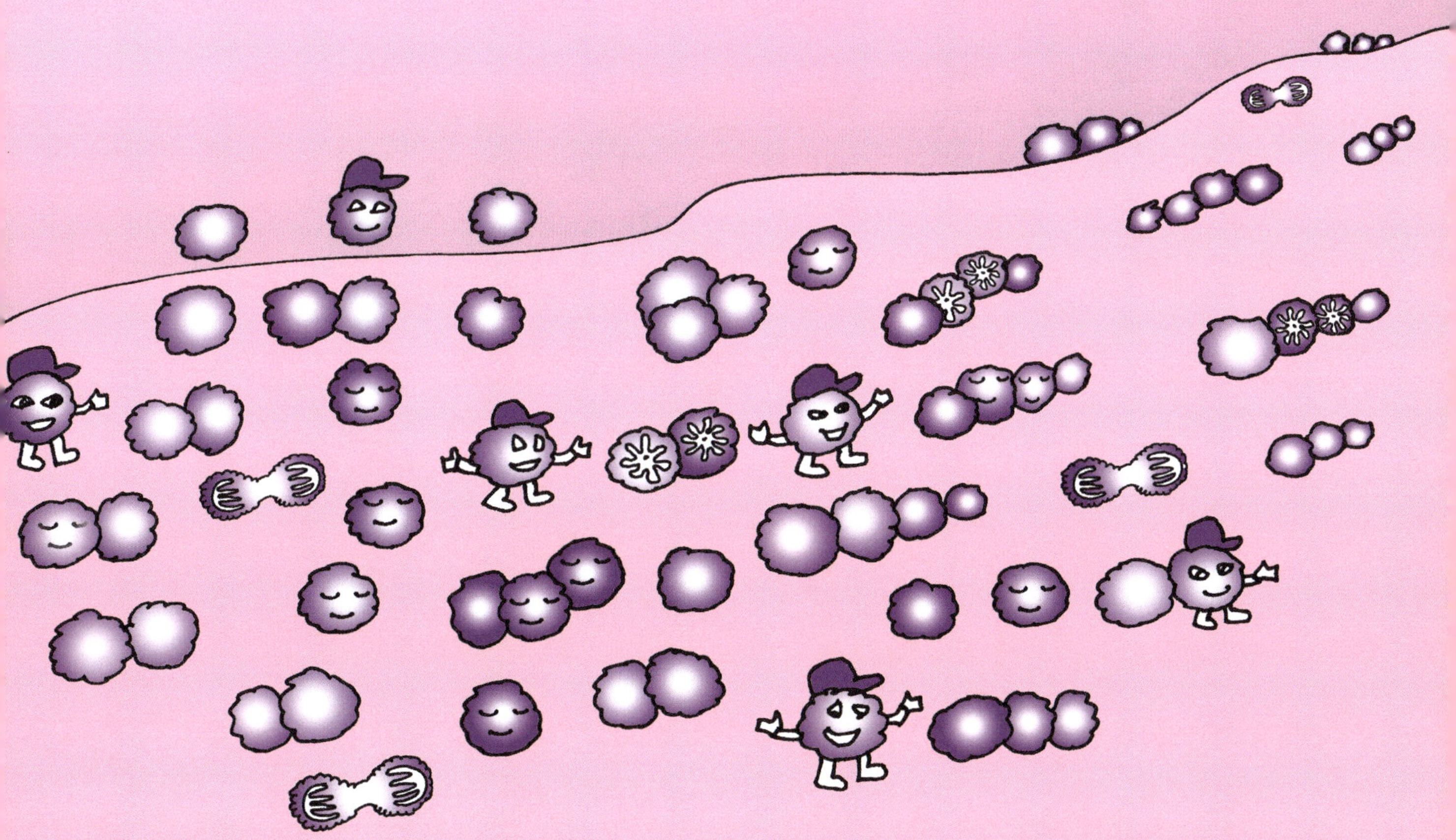

We send our love to injured cells
who struggle just to breathe —
we stay until they're stronger,
'til your cells no longer grieve."

Ricky dreamed about his cells all night
and their work to make him stronger.
They were small but they were loving –
and he wasn't sick much longer!

Then every night Rick went to bed
without a fuss or fight,
he called on Stemmie and his friends,
and sent them loving light.

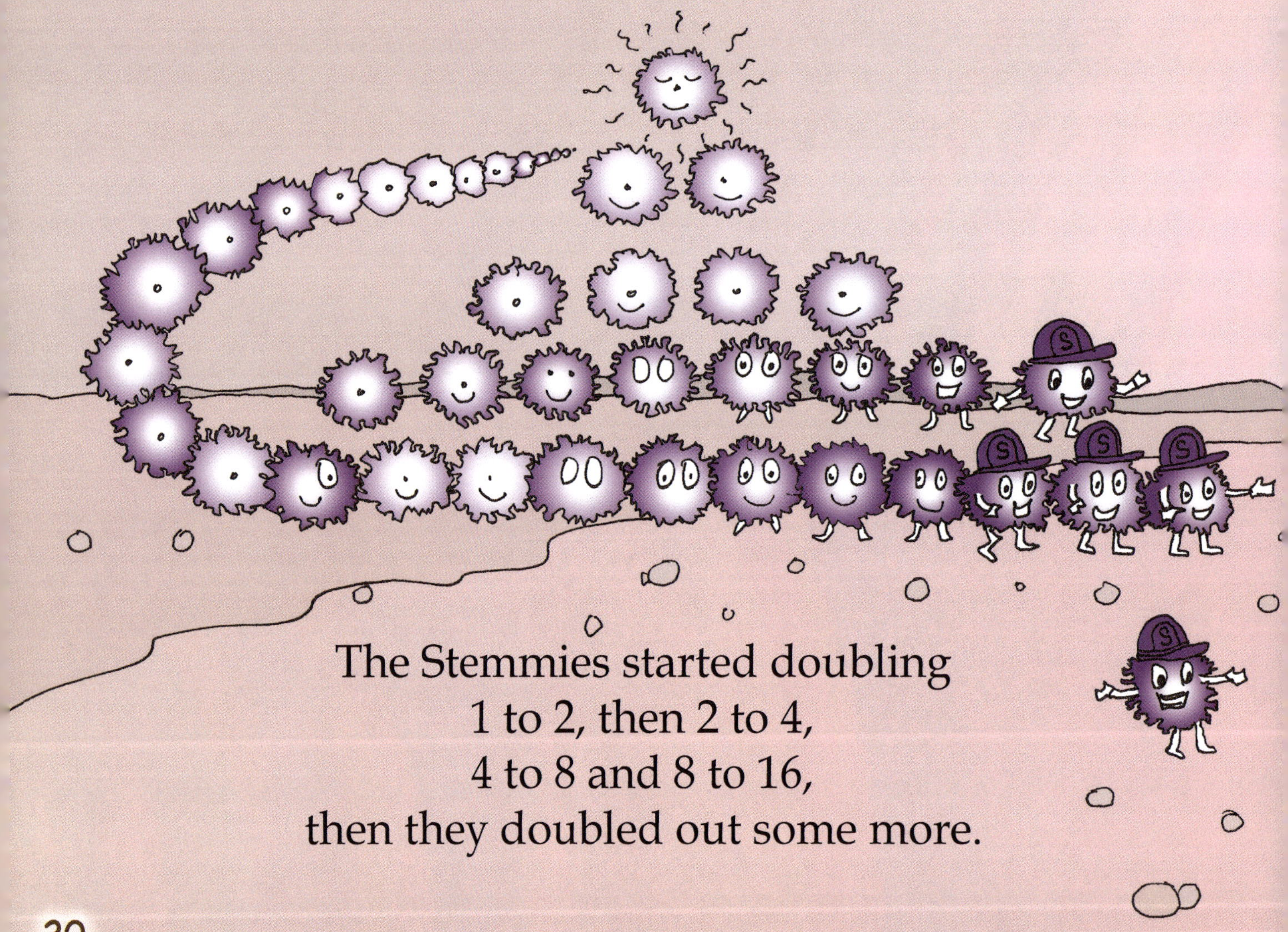

The Stemmies started doubling
1 to 2, then 2 to 4,
4 to 8 and 8 to 16,
then they doubled out some more.

The Mac cells grew in number
then quickly went to work
to seek out all the toxins and
where the bad guys lurk.

Some Stemmies fixed the leaks
in blood vessels that were hurt
while others made new brain cells
to help Ricky stay alert.

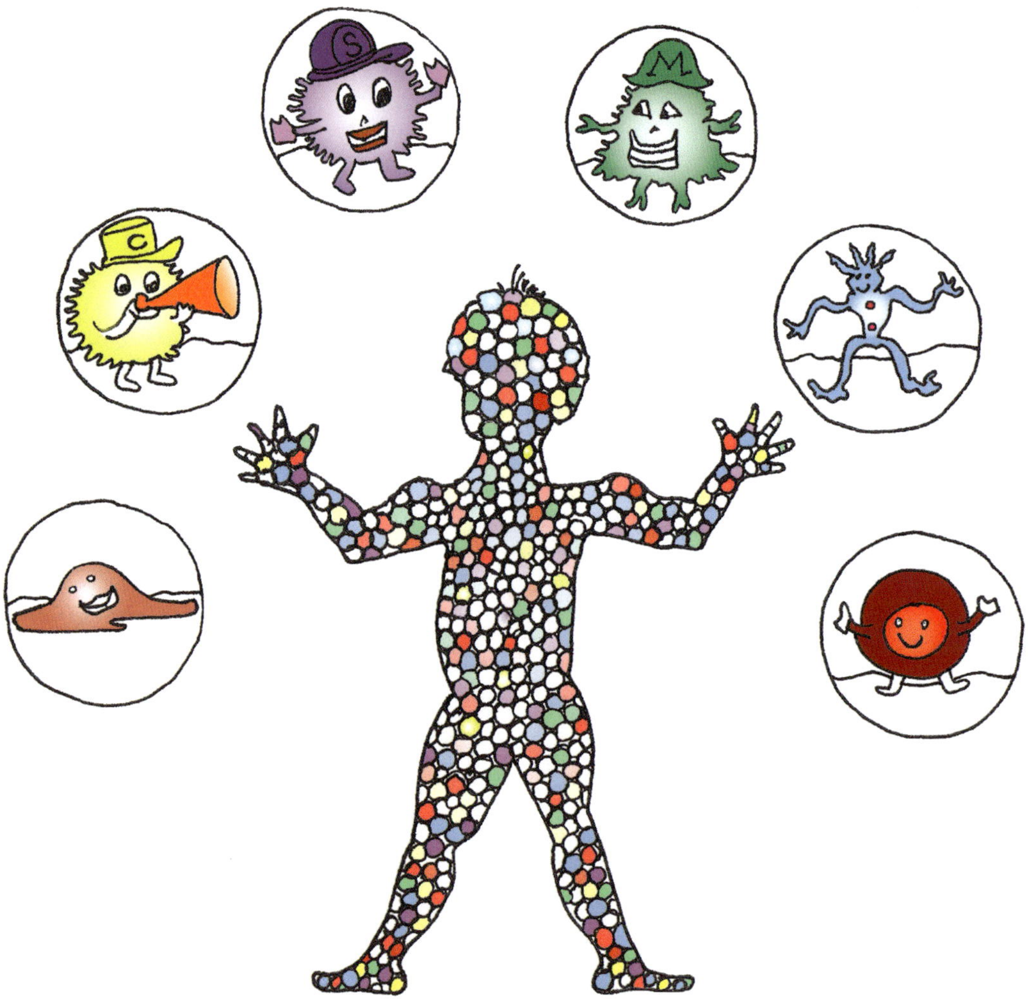

A few days passed and his parents knew
that something was going on.
Rick was eating veggies and
doing yoga with the dawn.

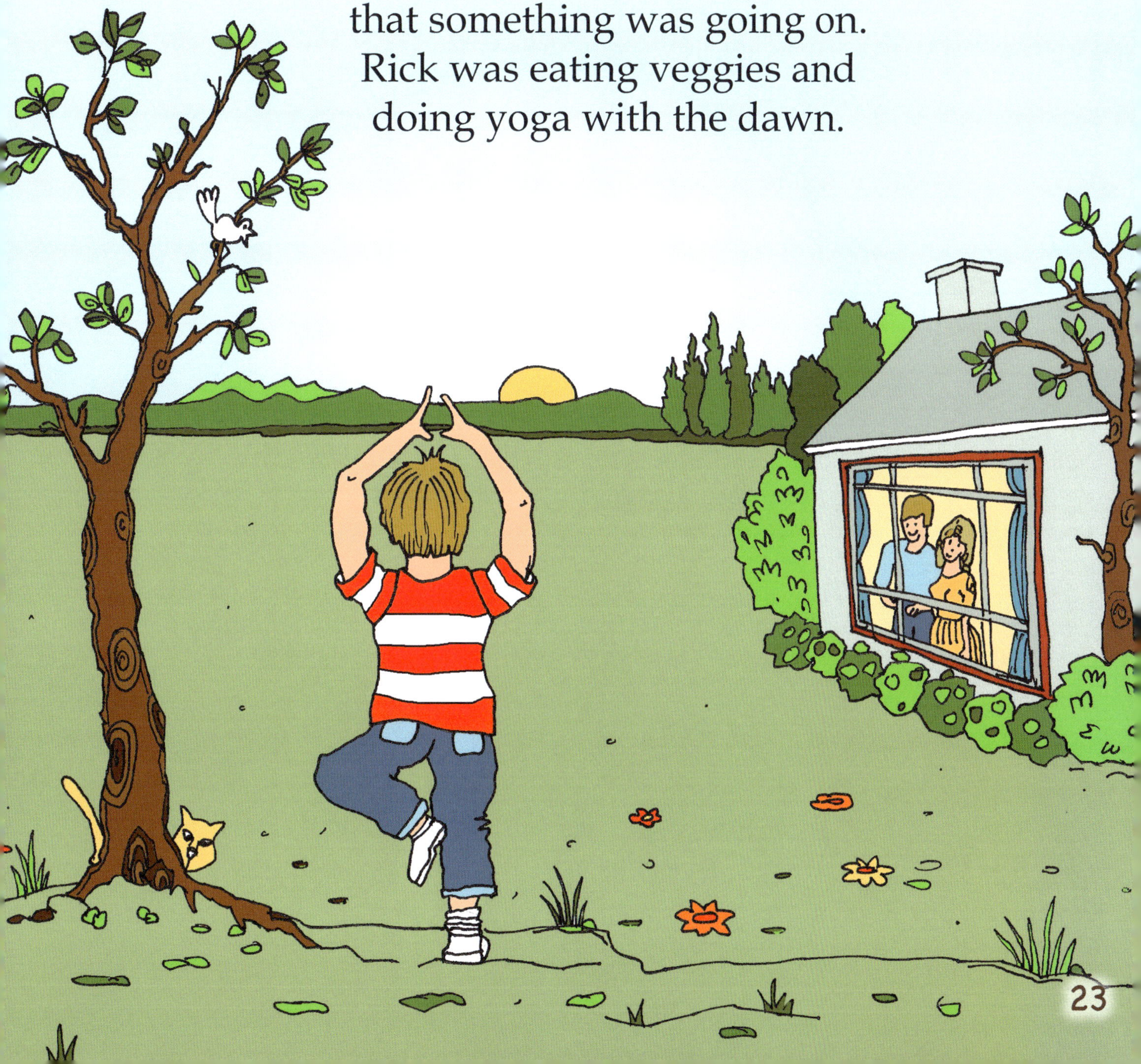

He looked at them with sparkling eyes
and told them about his dream…
the protecting cells, the healing cells,
all part of his wellness team.

"Stemmie also said
that I could be like him each day.
When I try to help my friends
and do my best at work and play.

Then he helped me think much bigger
about a larger team
where we care about each other
and share a common dream.

A dream that grows each minute
from 1 to 2, then 2 to 4,
from 4 to 8, then 8 to 16
then it doubles out some more
'til ...

...we're billions of Stemmies
renewing the planet
and healing each
silent scream,

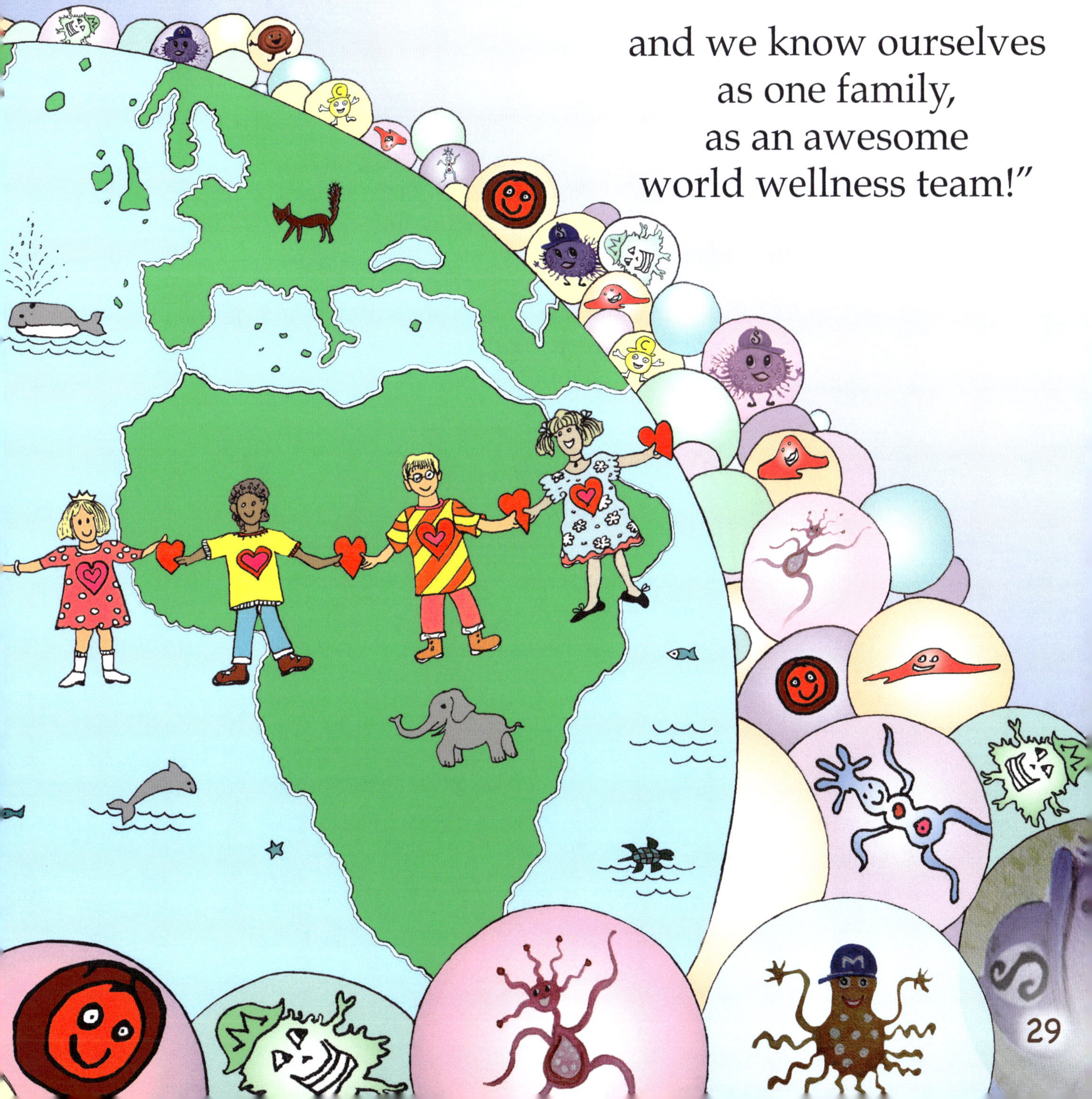

and we know ourselves
as one family,
as an awesome
world wellness team!"

GLOSSARY

Stemmie

A stem cell is like a parent cell that helps the other cells in your body when they get sick or injured. Some stem cells were the first cells in your body when you were a little baby. These Elder cells helped you develop your heart, brain, arms, legs, stomach and lungs – every part of you. After your birth, your stem cells have continued to help you grow bigger and stronger and smarter.

Muscle cells

Some of your stem cells can divide into muscle cells and help you grow stronger. Daily exercise is important to help keep your muscles in good shape. Moderate exercise also helps your heart muscle stay strong.

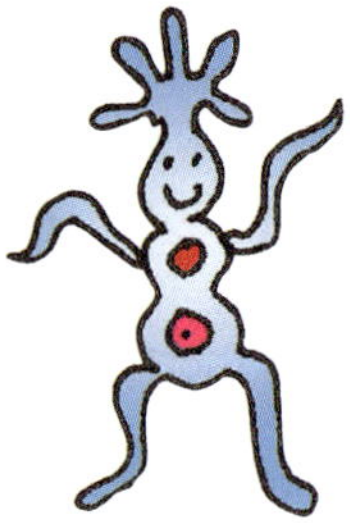

Neurons

Some of your stem cells divide into neurons for brain power. They help transfer directions from your brain to your arms and legs and organs. They help you learn new ideas and remember things from months and years ago. Neurons help you see, hear, smell, and taste and they help you feel a touch on your skin. They also help you talk, walk and plan fun things you would like to do.

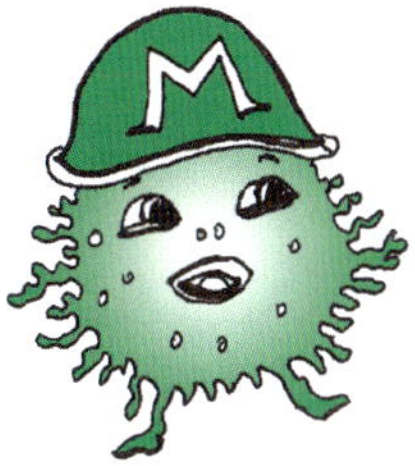

Mac

Mac is one of your transformer cells. Mac looks like a little immune cell (called a monocyte) until germs start hurting your cells. Then Mac grows bigger and bigger and then gobbles up anything that tries to hurt you.

Cytos

Cytos (called cytokines) are special messengers of your body. They help send out signals to the immune system that more cells are needed to defend your organs from being hurt by invaders like bacteria, viruses, chemical toxins, and other things that can hurt your cells.

Blood cells

Some of your stem cells divide to create your blood cells. These cells flow from your lungs and heart to other cells to give them oxygen power and good nutrition. The blood cells are also garbage collectors that take away the carbon dioxide waste. Blood cells are important for keeping all of the cells of your body healthy and happy.

REFERENCES

1. A stem cell can make identical copies of itself, produce a variety of different cells and repair and replace injured and dying cells. *Ahmed S., Journal of Cell Biochemistry, 2009; 106(1): 1-8.*

2. Negative thoughts can create stress and stress can reduce neural stem cells and their division into neurons. *Sapolsky, RM., Biological Psychiatry 2000; 48(8): 755-65.*

3. Exercise promotes growth factors that stimulate the growth of stem cells and new neurons. *Ploughman M., Developmental Neurorehabilitation 2008; 11(3): 236-40.*

4. Sleep deprivation reduces the growth of stem cells and new neurons. *Meerlo P. et al., Sleep Medicine Reviews 2009; 13(3): 187-94.*

5. Fresh natural foods provide nutrients that support stem cells. A high fat, refined sugar diet reduces neural stem cells, learning and memory. *Molteni R. et al., Neuroscience 2002; 112(4): 803-14.*

6. Air and water that are polluted with cadmium, arsenic or methylmercury are toxic to neural stem cells. *Watanabe J. et al., Journal of Toxicological Science 2009; 34(3): 315-25; Stummann TC. et al., Toxicology 2009; 257(3): 117-26.*

7. An enriched environment provides increased opportunities for learning and stimulates the growth of neural stem cells. *Seqovia G., et al., Journal of Neural Transmission, 2009; 116(8): 1007-16.*

8. Immune cells like monocytes and macrophages release little cytokine cells that are sent out to bring back other immune cells and stem cells. *Graber JJ., Dhib-Jalbut S., Pharmacology & Therapeutics 2009; 121(2):147-59.*

For further information on stem cells, visit our website at **http://www.auntielynny.com**

68009450R00022

Made in the USA
Charleston, SC
28 February 2017